Housing 124

失去光亮的眼睛

Seeing in a World Without Sight

Gunter Pauli

[比] 冈特·鲍利 著

[哥伦] 凯瑟琳娜·巴赫 绘

张晓蕾 唐继荣 译

上海远东出版社

丛书编委会

主　任：田成川

副主任：闫世东　林　玉

委　员：李原原　祝真旭　曾红鹰　靳增江　史国鹏
　　　　梁雅丽　孟小红　郑循如　陈　卫　任泽林
　　　　薛　梅　朱智翔　柳志清　冯　缨　齐晓江
　　　　朱习文　毕春萍　彭　勇

特别感谢以下热心人士对童书工作的支持：

匡志强　宋小华　解　东　厉　云　李　婧　庞英元
李　阳　梁婧婧　刘　丹　冯家宝　熊彩虹　罗淑怡
旷　婉　王靖雯　廖清州　王怡然　王　征　邵　杰
陈强林　陈　果　罗　佳　闫　艳　谢　露　张修博
陈梦竹　刘　灿　李　丹　郭　雯　戴　虹

目录

Contents

一只小螃蟹正在寻找食物，鼢鼠刚好经过，他看到这只甲壳动物那么小，还是全白的，感到很惊讶。

"哇，你有白化症吗？"鼢鼠问道。

"你真幸运，居然能看到我是全白的？"螃蟹回答道。

A small crab is searching for food when a mole rat passes by and is surprised to see this tiny, completely white crustacean.

"Oh, you are an albino?" Mole Rat asks.

"And, are you so fortunate that you can see I am all white?" the crab responds.

哇，你有白化症吗？

Oh, you are an albino?

……自从火山爆发之后……

... since the volcanoes erupted ...

"你看不到自己的样子吗？你周围没有镜子吗？"

"我看不到！自从三千年前这里火山爆发之后，我就一直生活在洞穴里。我不仅失去了视力，连眼睛也退化了。"

"因此你才被称为'白化盲龙虾'？"

"Can't you see yourself? Isn't there a mirror around?"

"No, and since the volcanoes erupted here three thousand years ago, I have been living in caves, so I have not only lost my sight, but also my eyes."

"So you are called the blind albino lobster?"

"好吧，我猜想最早见过我的那些人都没有戴眼镜，因为他们叫我'白化螃蟹'，而其实我是一只盲龙虾。"

　　"我倒是有两只眼睛，不过它们更多是用来装饰我的脸罢了，我几乎从来用不上它们。"

　　"你也生活在黑暗中吗？"

"Well, I think the people who saw me first did not have their glasses on, as they called me an albino crab, even though I am a blind lobster."

"I do have two eyes, but they are more for decorating my face as I hardly ever use them."

"Are you then living in the dark as well?"

你也生活在黑暗中吗？

Are you then living in the dark as well?

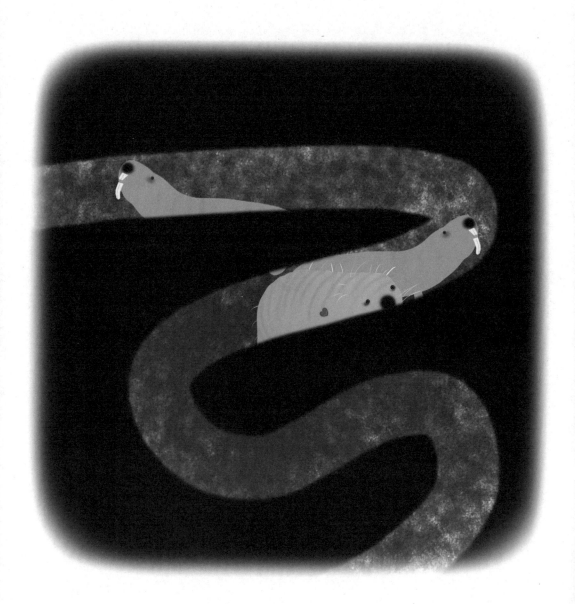

... fully equipped burrows underground ...

"没错，我生活在黑暗中。我们把洞穴建在地底下，样样齐全，有睡觉的房间、育儿室、厕所，甚至还有一个食品储藏室。"

"那你吃昆虫吗？"

"不吃！我是素食主义者，我们鼹鼠只吃树根和茎。"

"Indeed, I do. We build our burrows underground, fully equipped with sleeping quarters, nurseries, toilets and even a pantry."

"Do you eat insects?"

"No, I am a vegetarian. We mole rats live on roots and bulbs."

"你就这么生吃？一定很难消化！如果让我吃，我会消化不良的。"

　　"别为我担心。我们的肠道里有很多细菌，可以将所有这些地底美食转化为整个群体共用的营养物质。"

"You eat fresh, raw food? That must be hard to digest. If I were to eat it, I would have indigestion."
"No need to worry about me … We have many bacteria in our gut that will turn all these underground goodies into healthy nutrition for the whole colony."

……生吃食物？

... fresh, raw food?

······一位女王······

... a queen ...

"群体？你是说家族吗？"

"不，不是的！我们有一整个群体，里面有一位女王，她是巢穴中唯一的雌性，负责繁殖后代。群体中的其他个体来回奔忙，保护巢穴，并为每个个体提供食物。"

"Colony? You mean family?"
"No, no! We have a whole colony – one with a queen, and she is the only female in the nest to have litters of little ones. The rest of us run back and forth to defend and feed everyone."

"请原谅我的无知，请问你是鼹鼠吗？"

"啊哈！你被称为螃蟹，但你其实是龙虾。我虽然被称为裸鼢鼠，但与鼹鼠或大鼠相比，我跟豪猪和豚鼠类的血缘关系更近一些。"

"人们怎么会如此稀里糊涂呢？"

"Excuse my ignorance, but are you a mole?"

"Ah ha! You are called a crab, when in reality you are a lobster. I am called a naked mole rat, but I am closer to porcupines and guinea pigs than moles or rats."

"How can people get so confused?"

... closer to porcupines and guinea pigs ...

......我们身上用来感知周围环境的体毛。

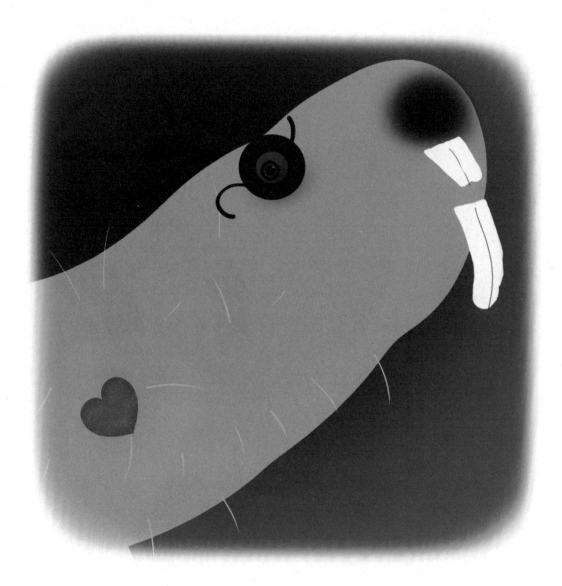

...hairs on our body, to feel our way around.

"不要责怪他们的无知。他们轻易相信自己所见到的，却很少花时间去研究与他们一样生活在地球上的每种生物的独特性。"

"那么与脂鲤和洞螈这两种盲视生物相比，你们有什么独一无二的地方吗？"

"我们都很长寿，也从来不会得癌症。而且，我们身上用来感知周围环境的体毛不到100根。"

"Do not blame them for their ignorance. They quickly believe what they see and seldom take the time to discover the uniqueness of every creature with whom they share the Earth."

"And what is so unique about you, when compared with the tetra and the olm, two creatures that cannot see either?"

"Well, we live long lives. We never get cancer. And we have no more than a hundred hairs on our body, to help feel our way around."

"所以你不仅看不见，还不长毛？"

"毛发和眼睛不是生命中最重要的东西，食物和家庭才是。就像你和其他所有生物一样，我们也想在自己的群落中活得长久、快乐和健康。"

"你说得对！保持简单的生活——尽管通常说比做更容易。"

……这仅仅是开始！……

"So you are blind and naked?"
"Hair and eyes are not the most important things in life – food and family are. Like you and everyone else, we too want to live long, happy and healthy lives in our precious community."
"You said it right, and keep it simple. Although it is often easier said than done."
... AND IT HAS ONLY JUST BEGUN! ...

......这仅仅是开始! ......

... AND IT HAS ONLY JUST BEGUN! ...

Did You Know?

你知道吗？

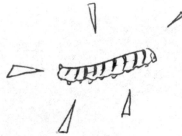 

动物和植物的命名经常出错。比如蚕并不是蠕虫，而是毛虫。蓟和荨麻被称为杂草，但它们实际上富含营养物质和生化物质。

Animals and plants are often given the wrong names. A silk worm is not a worm, but rather a caterpillar. Thistles and nettles are called weeds, while they are in fact rich in nutrients and biochemicals.

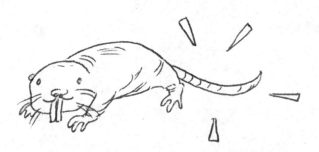

Mole rats have a tail like rats do, and dig burrows like moles do, but in fact mole rats are closely related to porcupines, guinea pigs and chinchillas.

鼢鼠有像大鼠一样的尾巴，它们也像鼹鼠一样打洞，但事实上鼢鼠跟豪猪、豚鼠和毛丝鼠的血缘关系更近。

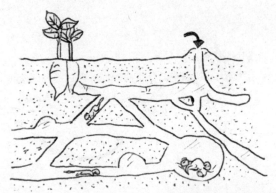

鼢鼠是唯一一种以和蚂蚁、蜜蜂一样的群体方式生活的哺乳动物。数量高达 300 只的鼢鼠个体几代同堂，都住在一个位于地下的"鼢鼠镇"，洞穴面积展开后有 6 个足球场那么大。

The mole rat is the only mammal that lives in a colony in the same way that ants and bees do. Several generations of up to 300 individuals live in an underground "mole town" that can stretch over an area the size of up to 6 soccer fields.

Mole rats can move their teeth like chopsticks. They do not need drinking water, as they remain hydrated with the moisture from their food.

鼢鼠像移动筷子一样移动它们的牙齿。它们不需要喝水，食物中的水分足够满足它们的需要。

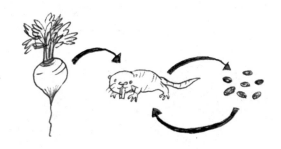

The roots and bulbs that mole rats eat are hard to digest, so in order to absorb more of the nutrients from that food, they eat their own droppings and digest the same biomass several times.

鼢鼠吃的根和茎都难以消化。为了从食物中吸收更多营养，它们会以自己的粪便为食，多次反复消化同一批生物质。

Mole rats not only eat their droppings, they also build a toilet chamber where they keep their waste and regularly roll in it so that all mole rats of that colony smell the same. As they can hardly see, this smell differentiates an intruder from a family member.

鼢鼠不仅以自己的粪便为食，还在巢穴里建造厕所间用来存放废物，并定期在里面打滚，所以同一个群体的鼢鼠身上都发出同样的气味。由于它们几乎看不见，这种气味能将入侵者与家族成员区分开来。

The queen mole rat can have a litter of 27 pups, and bears pups 4 to 5 times per year. After birth, pups are cleaned and taken to a nursery, where the mother visits them daily to feed them.

鼹鼠女王一次可以产仔 27 只，每年可以产仔 4 到 5 次。幼崽出生后被洗干净，然后送往育儿室。女王会每天来看望幼崽，给它们喂食。

4,000m

Blind albino crabs (that are in fact lobsters) live in caves up to 4,000 metres under sea level on the island of Lanzarote (Canary Islands, Spain). These crabs are endangered due to noise pollution, as they are affected by excessive noise.

西班牙加纳利群岛的兰萨罗特岛上的白化盲蟹（事实上属于龙虾）通常生活在海平面以下 4 000 米深的洞穴里。这些盲蟹易受超量的声音影响，它们因噪音污染而濒危。

Think About It

想一想

Why do we use a name for a plant or an animal that is obviously not correct?

为什么我们会给一种植物或动物明显不正确的命名呢？

If there were no light, and you were living in a dark world, of what use would your eyes be?

如果周围没有光照，你一直生活在一个黑暗的世界，你的眼睛还有什么作用？

Are appearance and good looks more important than family?

长相好看比家人更加重要吗？

Do you feel you are just one of many, or do you consider yourself special?

你觉得自己只是许多人中普通的一个，还是认为自己独一无二？

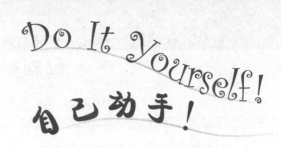
Close your eyes. Now sense the world around you. Feel the chair you are sitting on, or the desk you are leaning on. If you are at home, find your way to the kitchen, with your eyes closed. If you need to, open you eyes briefly to orientate yourself before moving. Now go and stand in front of the refrigerator with closed eyes, open the door and find different jars, the juice bottle, the butter and the eggs. Do you think you will be able to find everything you need to prepare a simple slice of bread with jam? If you are at school, get some friends together and go out to the garden or playing ground. Close your eyes and count the number of steps you are able to take confidently without opening your eyes. Now talk to your friends and find out who is able to take the most steps with eyes closed, without bumping into something or someone else.

闭上眼睛。现在来感知你周围的世界。感受坐着的椅子，或倚靠的桌子。如果在家，就闭上眼睛，摸索着找到厨房。如果有必要，可以短暂睁开眼，以便在迈步前确定方向。现在闭着眼睛继续向前，找到冰箱并站在它的前面；然后打开冰箱门，找出不同的罐子、果汁瓶、黄油和鸡蛋。你认为你能找到所有原料，做一片简单的果酱面包吗？如果在学校，找些朋友一起去花园或操场玩耍。闭上眼睛走路，数一数你在没有睁眼的情况下能自信地走多少步。邀请你的朋友一起参与，看看谁是在闭眼且不撞到障碍物的前提下，走出步数最多的人。

学科知识
Academic Knowledge

生物学	眼睛的生物学——虹膜、晶状体、瞳孔、视网膜、视神经、玻璃状液；鼹鼠是哺乳动物中唯一不能自主调解体温的；鳟、蛤蜊和海参都是令人吃惊的穴居者；鼹鼠的核糖体在转录蛋白质时不会有任何差错，所以鼹鼠不会得癌症；鼹鼠是最长寿的啮齿动物（寿命可达30年）；水螅是一种小型水母，通过触须末端上的光感蛋白来"看"周围的环境。
化 学	眼泪里含有水、油、黏液；白化体的眼睛、皮肤和头发都缺少色素；生鲜食物中抗氧化剂番茄红素的含量较低；另一些蔬菜，如芦笋、卷心菜、胡萝卜、蘑菇、胡椒、菠菜等蒸熟或煮熟后比生吃能为身体提供更多的类胡萝卜素和阿魏酸之类的抗氧化剂；裸鼹鼠的新陈代谢速率最多可降25%。
物 理	视网膜上有上百万个微小的感光神经细胞，被称为视杆细胞和视锥细胞；视杆细胞负责夜视（暗光）、外周或侧视，并检测运动；视锥细胞则专门感知颜色和细节；眼泪既能够润滑眼球表面又能冲走碎屑；烹制含有番茄红素的食物（如西红柿），能显著提高抗氧化剂的生物可利用性；有些食物生吃比熟食更好，因为这些食物含有的生物光子光能和酶会被高温破坏；裸鼹鼠的皮肤缺少一种哺乳动物负责向中枢神经系统发送疼痛信号的关键神经递质，因而它们不会感到疼痛。
工程学	动物洞穴、地下避难所、产仔的巢穴或食物存储系统的建设。
经济学	动物洞穴会威胁人类农业和住宅发展；改善全社会盲人的识字状况有利于提高就业率。
伦理学	举一个你知道的错误的命名。
历 史	视力大约出现在5亿年前；路易·布莱叶在1837年出版了包括音乐符号在内的第一部二进制形式的作品《点字盲文》。
地 理	裸鼹鼠居住在非洲东部；生活在美国大平原的黑尾草原犬鼠最大的洞穴系统面积可达6.4万平方千米，里面有多达40万只黑尾草原犬鼠。
数 学	一个完整的盲人点字单元有排成两行的六个凸起的点，其中每行三个点，当使用一个或多个点的时候，最多可以实现64种排列方式。
生活方式	烹饪过的食物能够给大脑中大量的神经元提供营养；裸鼹鼠如此长寿，是因为它们在严酷的环境下可以减慢自身新陈代谢，预防氧化应激；我们会相信眼睛所见和道听途说，但很少花时间去为自己发掘真相。
社会学	社会凝聚力能让社会度过艰难困苦，并在太平盛世提供更多欢乐。
心理学	负面的家庭关系会导致压力激增，影响心理健康，甚至造成身体疾病；幸福快乐源于大家一起做事以及为他人做好事。
系统论	草原犬鼠的食料（主要是草）促进向日葵和苜蓿的生长，而丰富的植物吸引了美洲野牛，也为家畜提供更多营养丰富的饲料。

情感智慧
Emotional Intelligence

裸鼢鼠

裸鼢鼠率真而又自信地问出一个明显涉及隐私的问题。他谈论起身体的重要特征和彼此的外表，似乎这都无所谓，自己的失明也不值得悲伤。他骄傲地向白化蟹介绍他们精心构建的洞穴系统。谈到饮食问题，他坦率地表明他们生吃食物。他知道这将引起有关营养物质的关注，他对批评早有准备，指出他们与肠道微生物紧密的共生关系，并且确信这样能组成一个有效的社会组织。他对自己的出身来历感到自豪，因为他与鼹鼠和大鼠完全不同。关于是什么使他如此独特，他有一个现成的答案：知道如何从生活的琐碎细节中识别那些重要的事情。

白化穴蟹（龙虾）

这只龙虾含蓄地用问题回答问题，传达了"你可以看到我，但我看不到自己"的意思，不动声色地解释了自己的创伤。他拿自己被错误命名的事开玩笑，说人们最初发现他的时候没有看太清楚。龙虾表现出对鼢鼠的同情，但把交谈的主题从视力转移到住房和食物上。他力求正确地理解，询问那些人们从来都懒得问的细节问题，因此他对人们如此稀里糊涂地就给出错误的命名和描述感到惊讶。鼢鼠提供了答案：人们不会花时间去了解所有生物的特性。当龙虾打听鼢鼠的外表（看不见还不长毛）时，他立即受到了鼢鼠关于什么是生活和幸福的教育。

艺术
The Arts

盲文是一门艺术。盲人看不见，有没有办法让他们也有机会欣赏艺术？探索一切可能，去创造一件盲人能欣赏的艺术作品。找几位朋友一起完成这项任务，并提出一项建议，让视觉受损的人更容易接触到艺术。与父母和老师一起讨论你提出的艺术项目，然后再创建和分享你的计划。

思维拓展
Systems: Making the Connections

视觉是进化发展最晚的感官中的一种。根据英国科学家安德鲁·帕克的说法，眼睛的出现触发了物种的巨大发展。生物现在可以看到它们的捕食者，也可以被捕食者看到，因此猎物不得不采用伪装技术，而捕食者也得调整它们的捕猎技巧。然后是生物多样性大爆发，进而引发从装甲盾牌到生物发光、360度视野、隐形蛛网或回声定位等最有创意的反应。在地球的历史上，这样的发展似乎空前未有。虽然眼睛在没有终点的进化过程中出现，现在又成了一种占主导地位的感官，但还是有一些情况下眼睛用处不大。所以，尽管视力几乎对所有陆地和海洋生物来说都是突破性的进化结果，但对少数生物，例如洞螈和脂鲤或是在这则故事中描述的两种生物，视力因为缺乏用处而消退。一些物种通过进化适应了视觉这种能力，也有许多物种由于意外或疾病，只得生活在没有视力的世界里。世界上约有3 900万人一出生就没有或失去视力，另有2.5亿人的视力非常差。为克服视力的局限，他们发展出非凡的能力，表露出满足和快乐，在挑战困难的同时享受充实的时光。人们不仅发明了盲文的书写和阅读方法，还用计算机辅助的音频服务为沟通创造出新的选择。替代形式的多元和表达方式的创新，使人们认识到身体上的缺陷并不是通往幸福之路的障碍。归根结底，最重要的是有水、食物、居所和家人，而在满足这些条件的环境下，每个人都可以在他们的社群里过上长久、快乐和健康的生活。

动手能力
Capacity to Implement

根据拉美人的传统，在孩子生日时，哥哥、父亲或教父会做一个皮纳塔——一个由硬纸板做成的大雕像，里面装着糖果和玩具，悬挂在花园里，让孩子们蒙上眼睛用棍子击打。在多次击打（失手）之后，皮纳塔被打破，散落出让人惊喜的礼物。看看不同的皮纳塔设计，并为下一个生日聚会做一个皮纳塔。大家都希望能在蒙着眼睛的时候击中皮纳塔，这个游戏会成为一个人人都想效仿的传统。也许你可以开一家皮纳塔生产公司！

故事灵感来自
This Fable Is Inspired by

萨布利亚·坦贝肯
Sabriye Tenberken

　　萨布利亚·坦贝肯出生于德国，在 12 岁时患上了一种疾病导致她失明。萨布利亚学习了哲学和社会学。在第一次访问西藏后，为西藏盲人创造了盲文。她和来自荷兰的保罗·克朗宁贝格建立了"盲文无国界组织"。她的书和电影获得了许多奖项。2006 年，她获得了中国政府颁发的"国际友谊奖"，因而获得广泛认可。现在，她住在印度喀拉拉邦，在那里为盲人继续奉献力量。

图书在版编目（CIP）数据

冈特生态童书.第四辑：修订版：全36册：汉英对照 /
（比）冈特·鲍利著；（哥伦）凯瑟琳娜·巴赫绘；
何家振等译.—上海：上海远东出版社，2023
书名原文：Gunter's Fables
ISBN 978-7-5476-1931-5

Ⅰ.①冈… Ⅱ.①冈… ②凯… ③何… Ⅲ.①生态环
境–环境保护–儿童读物—汉、英 Ⅳ.①X171.1-49

中国国家版本馆CIP数据核字（2023）第120983号
著作权合同登记号图字09-2023-0612号

策　　划 张　蓉
责任编辑 张君钦
封面设计 魏　来 李　廉

冈特生态童书
失去光亮的眼睛
[比]冈特·鲍利　著
[哥伦]凯瑟琳娜·巴赫　绘
张晓蕾　唐继荣　译

记得要和身边的小朋友分享环保知识哦！
八喜冰淇淋祝你成为环保小使者！